Book 2

Spatial
Relations

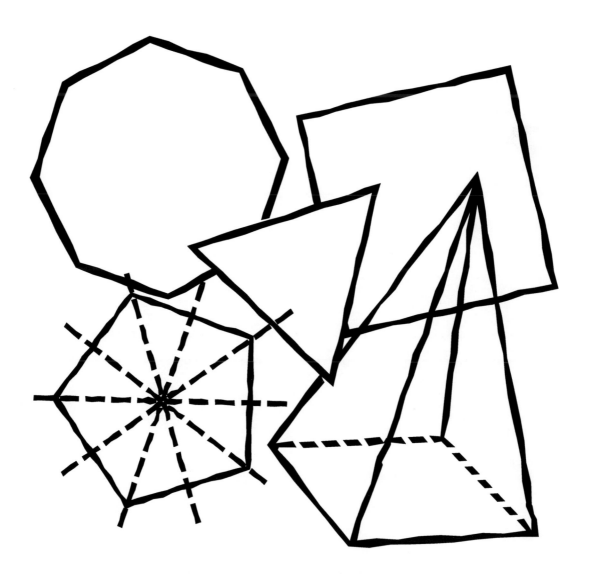

Written by David Thomas
Published by World Teachers Press®

Order Number 2-5104
ISBN 1-58324-026-8

C D E F G H 06 05 04 03

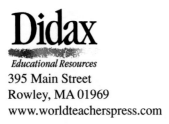

Didax

Educational Resources
395 Main Street
Rowley, MA 01969
www.worldteacherspress.com

Foreword

Spatial Relations is a three-book series covering the area of mathematics which relates specifically to space.

Spatial Relations Grades 3-4 provides ideas for work on:
- identification and description of shapes;
- their construction;
- pattern;
- simple symmetry;
- position;
- movement; and
- transformations.

Each section offers worksheets at various levels of difficulty, providing stimulating and informative activities for students. Much of the material is also suitable for use as extension or remediation in the desired areas.

The publisher has chosen to use metric measurements for most activities in this book. The National Council of Teachers of Math supports the use of the metric system as an integral part of the mathematics curriculum at all levels of education (NCTM Position Statement on Metrication, 1986). In some activities Imperial (English) measurements are used for illustrative purposes.

Contents

Teacher Information

Introduction

Spatial Relations is a vital part of any mathematics curriculum as it has an integral role in nearly all forms of math and also plays a major influence in our daily lives. As we perform various functions throughout the day, we require the concepts of space to help us through. "Please pass me the butter on the counter to the left of the refrigerator." This statement alone, requires the understanding of two spatial concepts. Imagine, how many times each day we have to rely on our spatial knowledge.

It is important to make students aware of the different forms of space. Often, an overemphasis can be placed on position in the environment which provides an unbalanced concept of space with students. Outlined below are the main forms of space, all of which are extensively covered in this series of books.

The ability of students to understand the concepts of space is developmental. For example, the concept of position is understood before the concept of grid coordinates and so on. Space is best tackled from the area of environmental awareness, as in most cases we are talking about spatial situations in our daily surroundings or events.

Movement and Position in the Environment

This section of *Spatial Relations* is involved with developing the understanding and interest of:
- movement and position;
- regions and networks; and
- representing objects with models and diagrams.

This is often the first area of space students are able to assimilate. This area of space is often quite straightforward to students, as they are already familiar with a great deal of the language involved. The language used is common to the home, after-school activities and general day-to-day experiences.

Exploring Two-dimensional Shape

This area of space concentrates on developing:
- familiarization with the attributes of shape;
- comparison and classification of shapes;
- representation by models and diagrams; and
- investigation of 2-D shapes in the environment.

The concepts dealt with in this area of space are relatively easy for students to understand, as this section requires a great deal of manipulation. It is very visual and hands-on.

Exploring Three-dimensional Shape

This area of *Spatial Relations* concentrates on developing:
- familiarization with the attributes of shape;
- comparison and classification of shapes;
- representation by models and diagrams; and
- investigation of 3-D shapes in the environment.

This section of the *Spatial Relations* curriculum is a little more difficult, as concepts are more abstract. Often activities are completed on paper, which means students have to transfer ideas from their heads to paper. This transferring of information is challenging for some students, and while they may have the knowledge, they may need more work in developing the skill to write or draw the information to reflect their knowledge and understanding.

Transforming Shapes

This section of *Spatial Relations* is involved with developing students' understanding and interest in:
- reflection, translation and rotation; and
- reduction, enlargement and distortion.

Transformation of shapes can take many forms and it is suggested that students begin with manipulation of materials before moving to pencil and paper activities. Some students may find it difficult to transfer their knowledge of these concepts from hands-on manipulation to that of recording with pencil and paper.

The following is a lesson development using one of the pages in this book. It is an example of how the activity could be introduced, developed and extended.

Activity: Pattern Making –1

Introductory Work

Introductory work at this level should center around discussing the ideas of patterns and space. Ask students to find examples in their environment. Bring examples to class to share in small groups. Discuss common elements, compare those elements that differ. Develop a bank of examples to help develop the knowledge to complete the activity.

Completing the Worksheets

The following is a suggestion for the development and extension of this activity.

1. Discuss the heading and develop a bank of examples from within the classroom where patterns are apparent.

2. Orally work through the worksheet with the students to ensure everyone is clear about what they are required to do.

3. Students complete the worksheet as per the instructions.

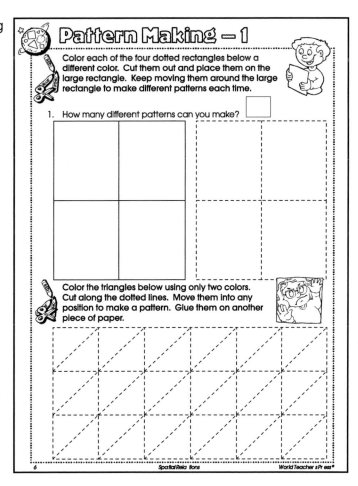

4. Students check their own work to ensure that it is complete and as accurate as possible.

5. Discuss the worksheet as a class group. Compare students' answers to find the variety of solutions. All students should be encouraged with positive feedback.

6. Collect worksheets for marking and monitoring of each student's development and understanding of the concept.

Extension

The extension of the activity is largely covered by the activities which follow, however, further discussion and development of patterns and space can occur if students are asked to extend the range of examples they already have. Encourage students to develop the concept further by finding obscure patterns in their environment or on creatures found in the environment.

Pattern Making — 1

Color each of the four dotted rectangles below a different color. Cut them out and place them on the large rectangle. Keep moving them around the large rectangle to make different patterns each time.

1. How many different patterns can you make?

Color the triangles below using only two colors. Cut along the dotted lines. Move them into any position to make a pattern. Glue them on another piece of paper.

Pattern Making – 2

Color these shapes to make an interesting pattern. Use only three colors. When you have finished, turn the page around and see how the pattern looks.

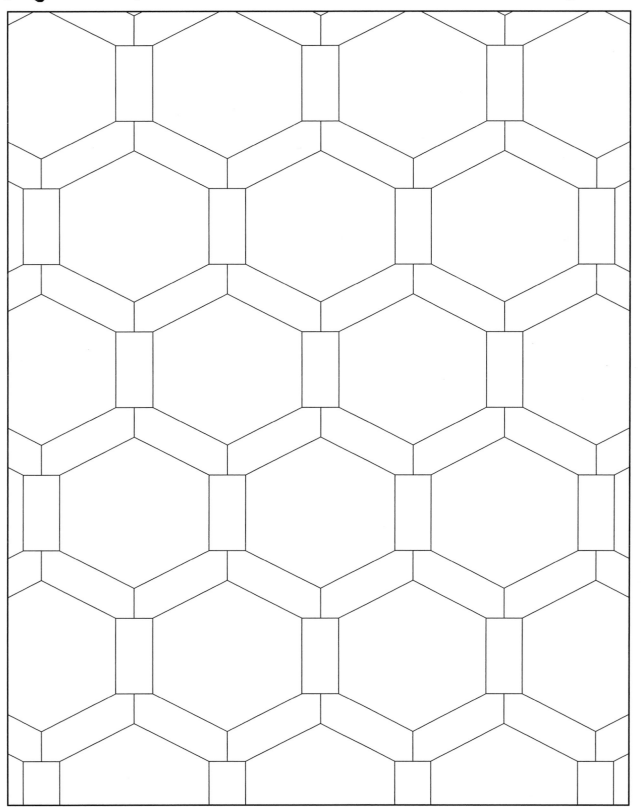

Instructions

In the box below, draw exactly what the instructions tell you.

1. A small red square in the top right-hand corner.

2. A blue pencil line from the bottom left-hand corner to the top right-hand corner.

3. A small green triangle in the bottom right-hand corner.

4. A yellow line from the bottom right-hand corner to the top left-hand corner, but stop when you get to the blue line. Now change the color of the line to brown.

5. Along the brown line, write your name in orange pencil.

6. Under the blue line, print your address in black pencil.

Football Stadium

This is Port City's football stadium.

Fred is sitting in row B, seat twenty-two.

	seat 20	seat 21		seat 23	seat 24
row B					

Look at this part of the stadium. Use the diagram to answer the questions below.

	seat 1		seat 3	seat 4	seat 5
row A		Bill			
	seat 11	seat 12	seat 13		seat 15
row B				Sue	
		seat 22	seat 23	seat 24	seat 25
row C	Tom				

1. Where is Bill sitting? _____

2. Where is Sue sitting? _____

3. Where is Tom sitting? _____

Color and cut out the pictures of Jill, Ken and Pat. Glue them on seats in the diagram above.

Jill

Ken

Pat

1. Where is Jill sitting? _____

2. Where is Ken sitting? _____

3. Where is Pat sitting? _____

4. What seat would you like to sit in? _____

Draw yourself in the seat you would most like to sit in.

Big Match

These are the seats at a football stadium.
Some are a different color and they spell the word CITY.

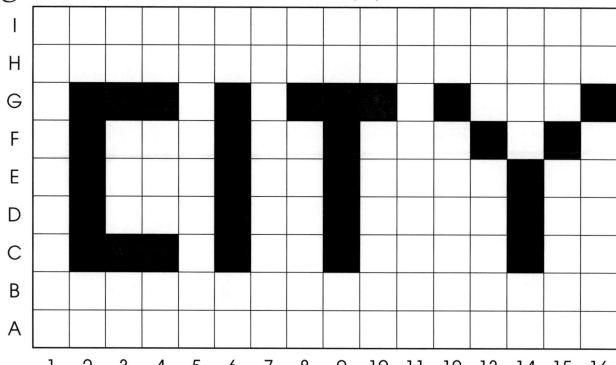

The seats (6,C), (6,D), (6,E), (6,F) and (6,G) spell the letter "I".

Which seats spell these letters…

1. C?_____

2. T?_____

3. Y?_____

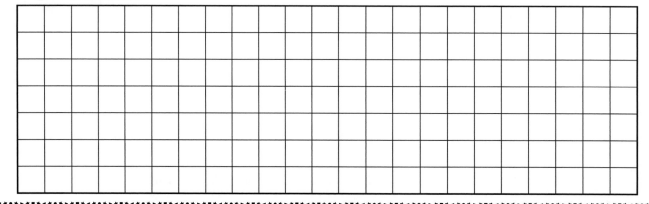

Color the seats which would spell the word "UNITED" at United's stadium.

Coordinate Points

Using the grid coordinates below, draw these things on your pirate island inside the grid squares.

1. Dead Man's Cave   (5,E)
2. Sword Creek   (5,F)
3. Danger Hill   (7,G)
4. Pirate Ship   (2,D)
5. Pirate Graveyard   (3,C)
6. Black Hills   (3,F)
7. Ghost Swamp   (7,D)
8. Shipwreck   (7,A)
9. Pirate Treasure   (8,G)
10. Landing Boat   (3,E)

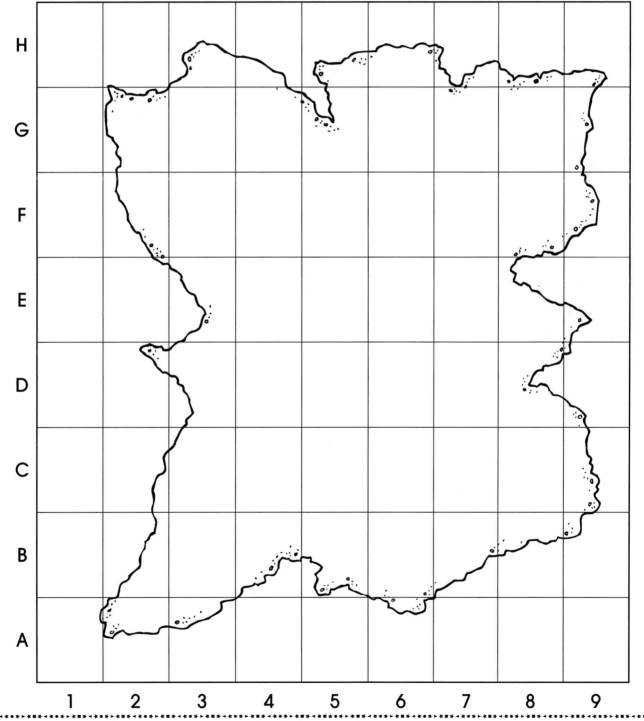

Grid Shapes

On the grid below, join the coordinates to draw a shape.

Write the name of the shape you have drawn. It will be one of these.

square ☐   rectangle ▭   rhombus ▱

parallelogram ▱   hexagon ⬡   triangle ◺

The first one is done for you.

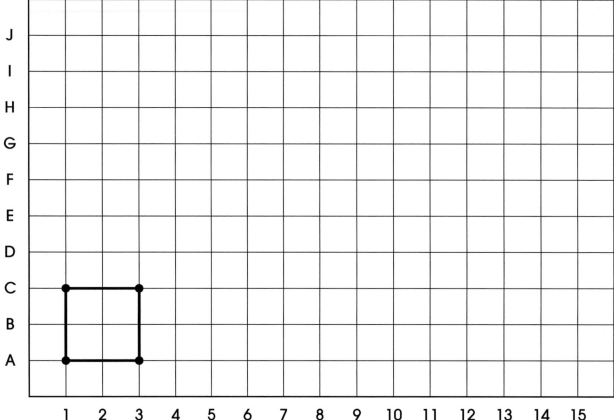

1. (1,A), (3,A), (3,C), (1,C), (1,A)   = square

2. (11, H), (11, J), (15, J), (15, H), (11, H) = _____

3. (1, H), (2, J), (4, J), (3, H), (1, H)   = _____

4. (2, E), (3, G), (7,G), (6, E), (2, E)   = _____

5. (10, B), (7, B), (10, E), (10, B)   = _____

6. (13, G), (14, G), (15, F), (15, E), (14, D),

    (13, D), (12, E), (12, F), (13, G)   = _____

Can you draw and name any more shapes?

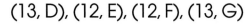

Line Design

Join the matching numbers with straight lines to see the interesting design you make. Color your design using two colors.

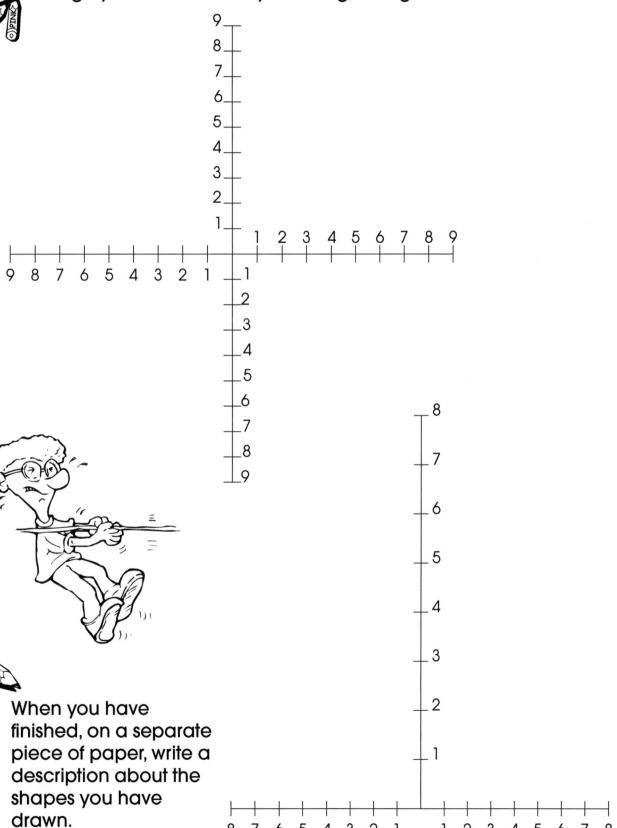

When you have finished, on a separate piece of paper, write a description about the shapes you have drawn.

Your Room

On the grid below, draw a plan of your classroom.
One-half a desk/table should equal one square.

What else could be included? Draw it on the grid.

Clockwise and Counterclockwise

This is a compass. The needle is pointing north. It also shows south, east and west.

If the needle (arrow) moves a quarter-turn clockwise, it will be pointing to the east.

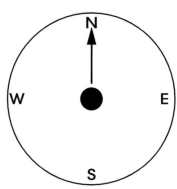

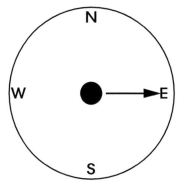

1. If the needle moves another quarter-turn clockwise, to which direction will it now point?

_____

2. If the needle now moves a half-turn clockwise, to which direction will it point?

_____

This compass starts with the needle pointing south.

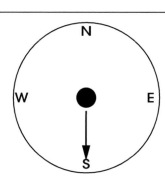

3. If the needle moves a half-turn clockwise, to which direction will it now be pointing?

_____

Draw the compass needle for these turns.

1. quarter-turn clockwise

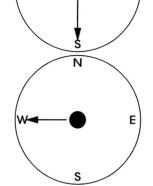

2. quarter-turn counterclockwise

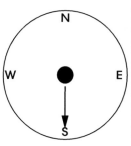

3. half-turn clockwise

4. half-turn counterclockwise

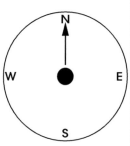

Fractions of a Turn

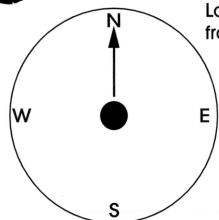

Look carefully at the needle positions on these fraction turns of the big compass needle.

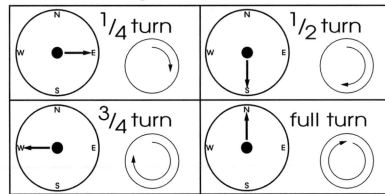

 Using the map below, follow the directions at each traffic circle to move to the next one. Start by facing north at the *school* and see which house you end at.

*Remember, at each stop, you will always face **north** before following the next direction.*

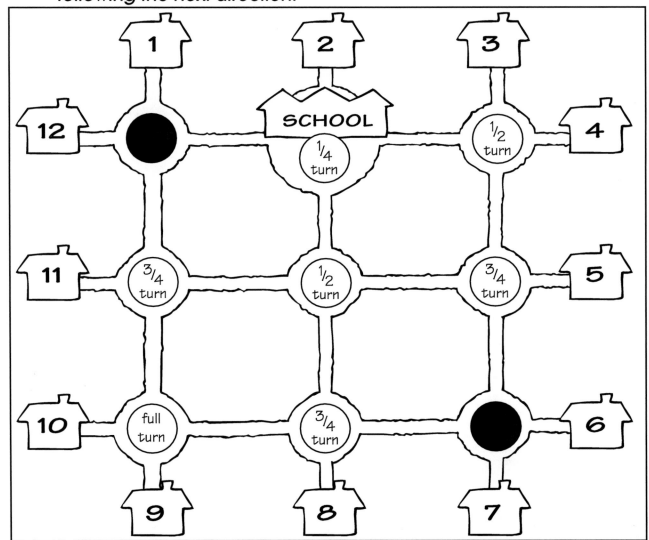

Finding a Way

Follow the directions from home **H** to find which store you are visiting.

*Remember, at each stop, you will face **north** again before following the next direction.*

Direction Turns

$^1/_4$ turn	→	move this way
$^1/_2$ turn	↓	move this way
$^3/_4$ turn	←	move this way
full turn	↑	move this way

1. Start at home **H**
2. $^1/_4$ turn, go two squares
3. full turn, go three squares
4. $^3/_4$ turn, go one square
5. full turn, go three squares
6. $^1/_4$ turn, go six squares

7. $^1/_2$ turn, go two squares
8. $^1/_4$ turn, go three squares
9. full turn, go four squares
10. $^1/_4$ turn, go two squares
11. full turn, go two squares

I am visiting the _____ store.

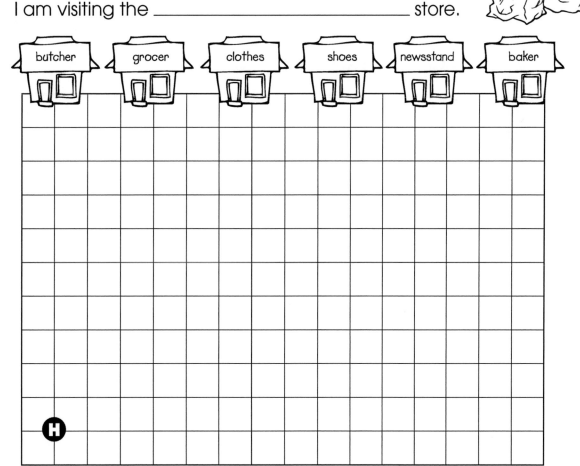

butcher   grocer   clothes   shoes   newsstand   baker

H

Use a colored pencil to draw a path to another store from home.
On a separate piece of paper, write the directions.

Compass

Follow the compass directions given and see what shape is formed. Use a colored pencil to mark the lines as you move.

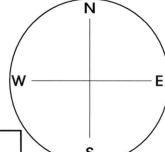

Directions: begin at "A"

N1, E1, N2, E2, N2, W2, N1, E3, S1, E3, S2, W3, S1, E4, S3, W3, S1, W1, N2, W2, S1, W1, N1, W1.

For example, N1 means north one square, S2 means south two squares.

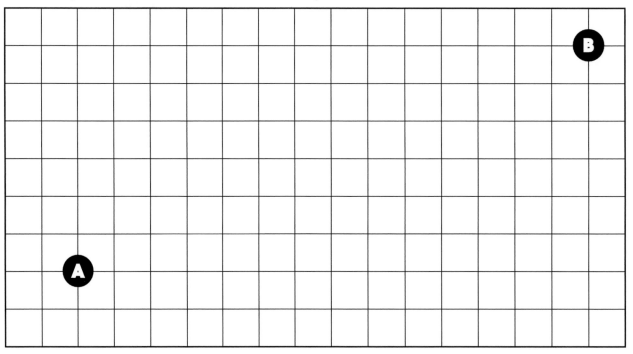

If you followed the directions correctly, you should have ended back at "A."

Find the area [ ] and the perimeter [ ] of the shape you have made.

Start at point "B." Make a rectangle. On a separate piece of paper, write the directions for your shape.

Find the area [ ] and the perimeter [ ] of the shape you have made.

Right Angles

When a door or window has opened a quarter-turn we can say it has made a right angle.

Just like quarter-turns, turning four right angles will take you back to where you started.

This is a right angle.

These are not right angles.

Circle the right angles.

1. 2. 3. 4. 5. 6.

How many right angles can you see in these shapes? The first one is done for you.

1. 2. 3. 4.

__4__ _____ _____ _____

5. Where can you see right angles in the classroom?

_____

_____

_____

_____

Chessboards

In the game of chess, pieces move in different ways.

The bishop (b) can move diagonally across squares like this:

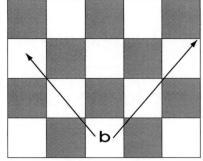

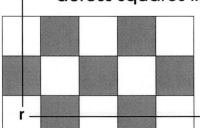

The rook (r) can move in straight lines horizontally or vertically like the diagram on the left.

On this chess board, show which squares the bishop could move to by putting a red circle in them.

Show which squares the rook could move to by coloring or outlining them blue.

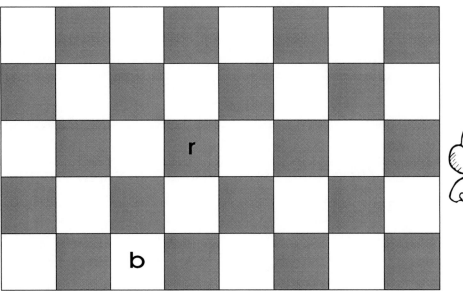

The queen (q) can move both diagonally and in straight lines. Put a queen anywhere on the board and color the squares she can move to, green.

Do you know how any other pieces move in chess?
Write their names below and describe how they move.

_____

_____

_____

_____

Pathways

Can you travel to all the towns on the map below without lifting your "pencil car" from the paper or going over the same road twice?

yes	no

Record the path you took below using a colored pencil.

Describe the path you followed.

I began at _____

_____

_____

_____

_____

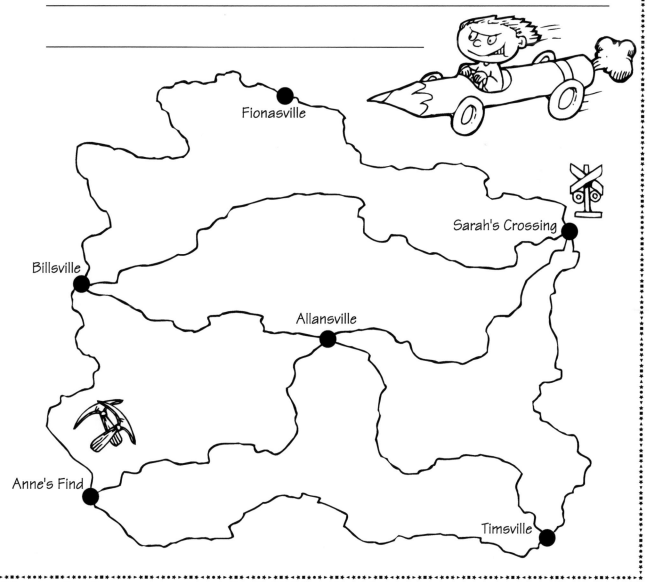

Reflecting Shapes

 If we put a mirror against this square…

the square would look like a rectangle.

 Put a mirror on the dotted lines and draw the shape you can see. Can you name the new shape?

1.

Name:

_____

2.

Name:

_____

3.

Name:

_____

4.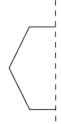

Name:

_____

Can you guess what this shape is?

Name:

_____

Squares and Rectangles

By flipping one shape over and over again it is possible to make more shapes.

flip ⟶

1. How many squares and rectangles can you find on this grid?

Squares [ ]   Rectangles [ ]

Find the area and perimeter of each shape.

Shape	Perimeter	Area	Shape	Perimeter	Area
1			2		
3			4		
5			6		
7			8		
9			10		

Enlargements

Double and triple the size of the shape drawn below.

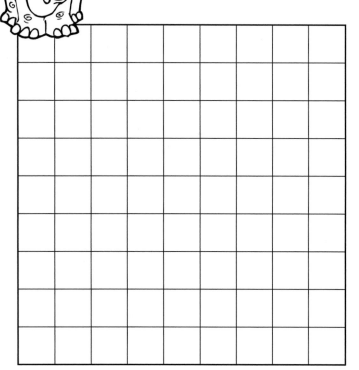

Original Size

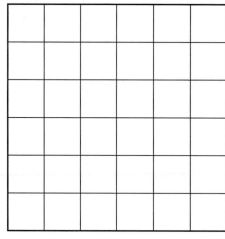

Double Size

Triple Size

Find the area and perimeter of each shape.

Shape	Perimeter	Area
Original Size		
Double Size		
Triple Size		

Triangular Grid

Shade the following shapes in different colors using the triangle grid above.

1. Three triangles of different sizes.
2. Three parallelograms of different sizes.
3. Three hexagons of different sizes.
 4. Three rhombuses of different sizes.

On a separate piece of paper, how many words can you make from the word parallelogram?

Patterns

The flower below has been drawn using seven different colors. Shade the petals using these colors. Turn the flower around one hexagon and draw it again. Shade the petals in their new positions. How many different flowers can you make?

Use the grid to show the new flowers.

Remember, the petals must stay in the same order.

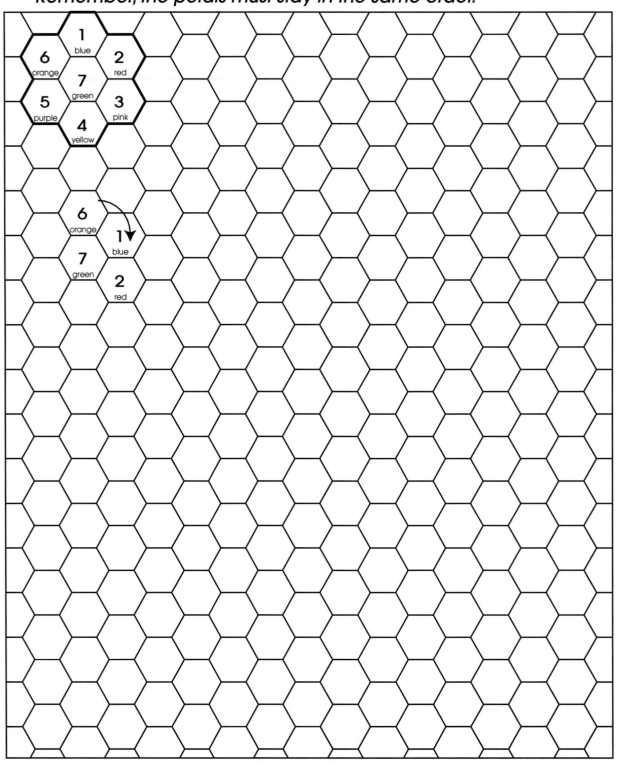

Covering Shapes

 Cover these shapes with counters.
Order them from smallest to largest.

A

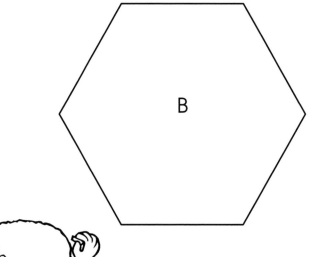
B

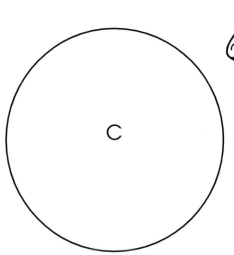
C

D

Shape	Number of counters	Order
A		
B		
C		
D		

Lines

In the picture below color all the horizontal lines blue, the vertical lines red and the sloping lines green.

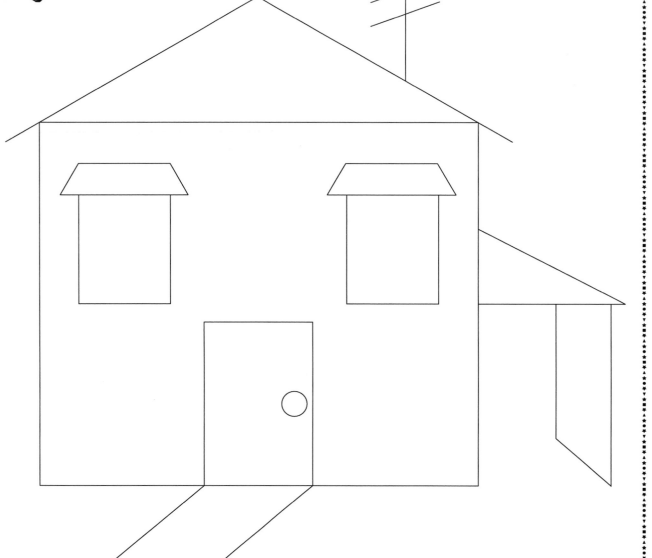

Complete the table below. Measure all the lines to find the total distance each color stretches.

Lines	Total Number	Total Distance
Red (vertical)		
Green (sloping)		
Blue (horizontal)		

Classifying Shapes – 1

Look at these shapes.

Aha! Round!

We could put them into two sets, or groups, like this:

(a)   – shapes with straight edges.

(b)   – shapes with curved edges.

Put these shapes into two sets and write down your reasons for separating them this way.

1.

(a) _____

(b) _____

2.

(a) _____

(b) _____

Classifying Shapes – 2

A B C D

E F G H

I J K L

Write all the shapes which have the following:
(remember, another name for side is edge)

1. three sides? _____

2. at least one right angle? _____

3. all sides the same length? _____

4. all sides different lengths? _____

5. four sides? _____

6. less than five sides? _____

7. no right angles? _____

8. at least one vertical line? _____

9. at least one horizontal line? _____

10. more than five sides? _____

Features of 2-D Shapes

Write the name of each shape, then complete the table below.

a. _____

b. _____

c. _____

d. _____

e. _____

How many vertices (corners) and edges (sides) does each shape have?

Do you notice anything about your answers?

_____

_____

_____

	vertices	edges
a		
b		
c		
d		
e		

What is different about these pairs of shapes?

_____

_____

_____

_____

_____

Describe one of the 2-D shapes in the box below to your partner. See if he or she can guess which one it is. Now it is your partner's turn to describe a shape for you to guess.

Use these words to help you: vertice (corner), side (edge) and angle.

a.

b.

c.

d.

e.

f.

g.

h.

Features of 3-D Shapes

 How many vertices (corners), faces (flat sides) and edges are there on these shapes?

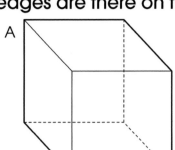

A

B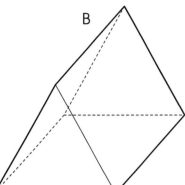

C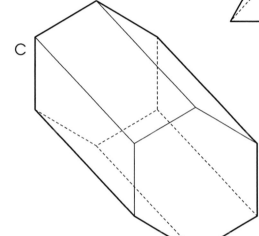

D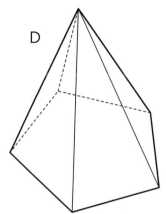

1. Complete the table.

Object	Vertices	Faces	Edges
A			
B			
C			
D			

2. Can you see a pattern in your answers?

_____

_____

_____

Guessing 3-D Shapes

Describe one of the 3-D shapes in the box below to your partner. See if he or she can guess which one it is. Now it is your partner's turn to describe a shape for you to guess.

Use these words to help you: vertice (corner), edge and face.

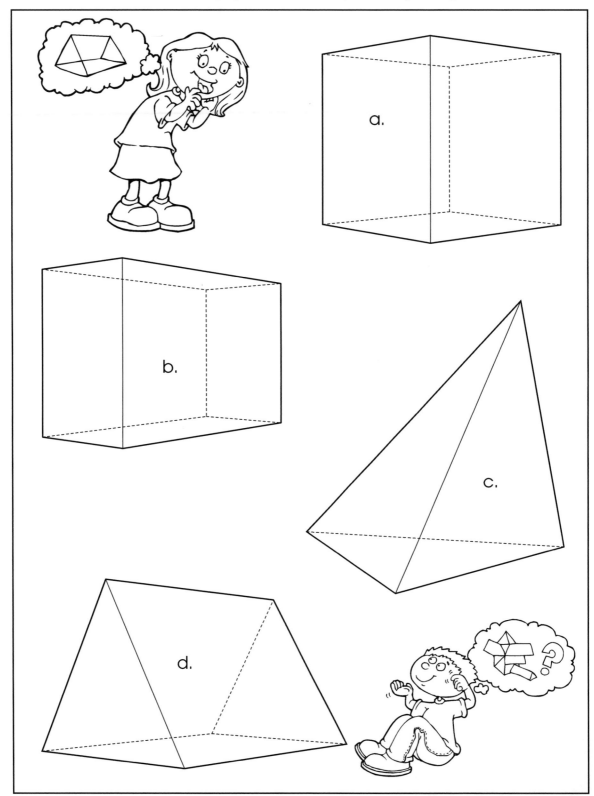

Spatial Relations Book 2 *World Teachers Press®*

Square Puzzle

Cut out the shapes below and put them together to make a square.

Is there more than one way to do this? _____

Triangle Puzzle

✂ Cut out the shapes below and put them together to make a triangle.

Is there more than one way to do this? _____

Tangram Pictures – 1

 Cut the tangram square along the dotted lines into its seven pieces and use these shapes to make the picture of the dog below.

dog

Tangram Pictures – 2

Use the tangram pieces from the previous activity to make these tangram pictures.

rectangle	triangle	parallelogram
buggy	yacht	cat
watering can	shoe	running person
swimming swan	mouse	teapot

Pentominoes

 The shape below is a pentomino. Pentominoes have... five squares, each joined along at least one edge. See if you can draw the other pentominoes. There are twelve altogether.

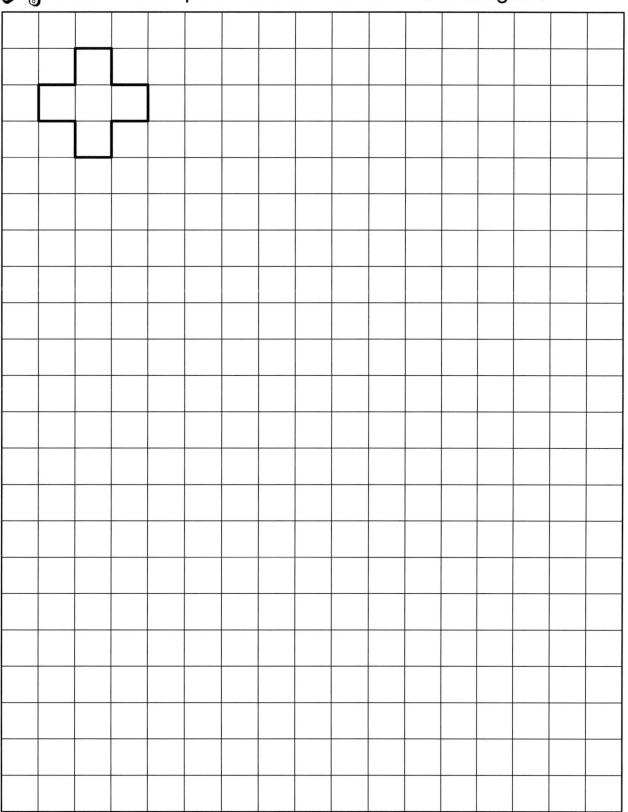

Looking at Cubes

Make these large cubes using smaller cubes.

A

B

C

1. Complete the table below.

Cube	Number of cubes	Number of faces	Number of corners	Number of edges
A				
B				
C				

Building with Cubes

Make these objects using smaller cubes.

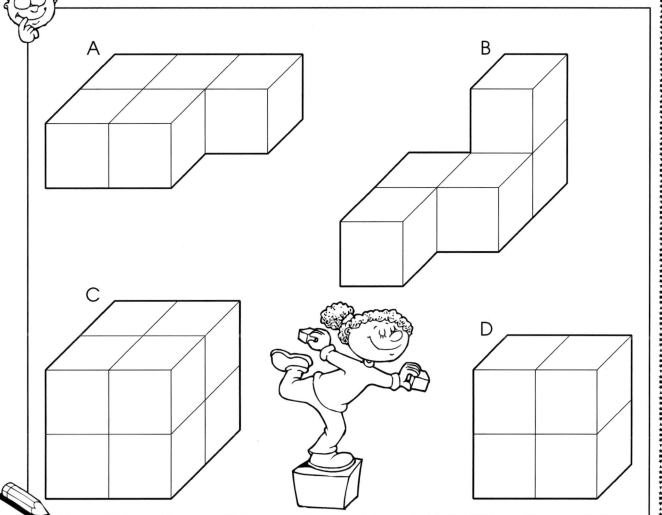

1. Complete the table below.

Shape	Number of cubes	Number of corners
A		
B		
C		
D		

2. Which shape contains the most number of cubes?_____

3. Which shape contains the least number of cubes?_____

4. Which shape has the greatest number of corners?_____

Objects and Shapes

 To make this 3-D shape you need these 2-D shapes.

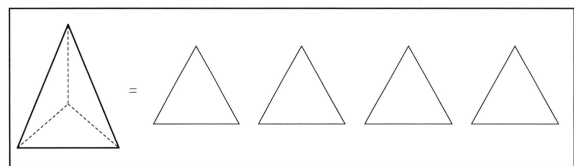

 Draw the 2-D shapes you need to make these 3-D shapes.

1.

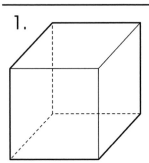

2.

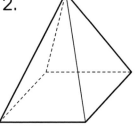

3.

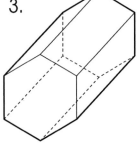

4.

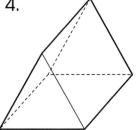

Building Objects

This shape can be built using six sticks or straws and four pieces of modeling clay.

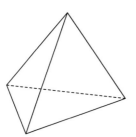

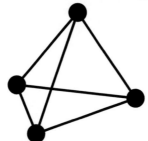

Try to make objects that have the following.

1. Nine sticks or straws and six pieces of modeling clay.

2. Twelve sticks or straws and eight pieces of modeling clay.

3. Eight sticks or straws and five pieces of modeling clay.

Constructing Objects

Make these objects using straws
or sticks and modeling clay.

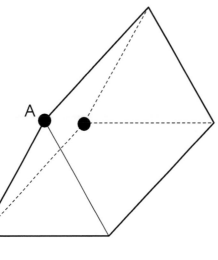

A

B

C

D

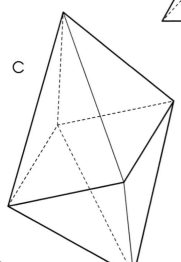

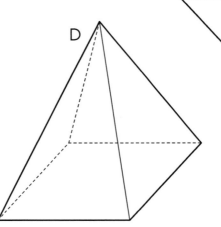

1. Complete this table.

Object	Pieces of modeling clay	Straws or sticks
A		
B		
C		
D		

2. Does the object with the most pieces of modeling
    clay use the most number of sticks or straws? _____

Tessellations

Two shapes can be used to make a tessellating pattern.
The grid below is made from octagons and squares.

Color the grid to make an interesting pattern.

Symmetry

Draw the mirror of these shapes to make numbers and letters. The first one is done for you.

1.

2.

3.

4.
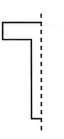

5.

6.

7.

8.

9.

10.

11.
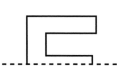

12.

13.

14.
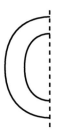

15.
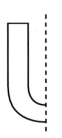

16.
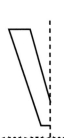

17.

18.

Answers

Page 6 – Pattern Making – 1
Teacher check

Page 7 – Pattern Making – 2
Teacher check

Page 8 – Instructions
Teacher check

Page 9 – Football Stadium
1. row A, seat 2   2. row B, seat 14
3. row C, seat 21
1. – 4. Teacher check
Teacher check

Page 10 – Big Match
1. (2,C), (2,D), (2,E), (2,F), (2,G), (3,C),
   (3,G), (4,C), (4,G)
2. (9,C), (9,D), (9,E), (9,F), (9,G), (8,G),
   (10,G)
3. (12,G), (13,F), (14,C), (14,D), (14,E),
   (15,F), (16,G)
Teacher check

Page 11 – Coordinate Points
Teacher check

Page 12 – Grid Shapes
square, rectangle, rhombus,
parallelogram, triangle, hexagon

Page 13 – Line Design
Teacher check

Page 14 – Your Room
Teacher check

Page 15 – Clockwise and Counterclockwise
1. south   2. north   3. north
1. west   2. east   3. east
4. south

Page 16 – Fractions of a Turn
house 11

Page 17 – Finding a Way
newsstand

Page 18 – Compass
A = 35, P = 44
Teacher check

Page 19 – Right Angles
Right angles = one, four and six
1. four   2. one   3. four   4. two
5. Teacher check

Page 20 – Chessboards
Teacher check

Page 21 – Pathways
Yes
Teacher check

Page 22 – Reflecting Shapes
1. triangle 2. rectangle
3. circle   4. hexagon
5. trapezoid

Page 23 – Squares and Rectangles
Squares = three, Rectangles = seven
1 – P = eight   A = four
2 – P = twelve   A = eight
3 – P = twelve   A = five
4 – P = eight   A = three
5 – P = fourteen   A = ten
6 – P = ten   A = four
7 – P = sixteen   A = sixteen
8 – P = ten   A = six
9 – P = fourteen   A = six
10 – P = twelve   A = nine

Page 24 – Enlargements
Original– P= twelve   A = five
Double– P = twenty-four   A = twenty
Triple   – P = thirty-six   A = fourty

Page 25 – Triangular Grid
Teacher check

Page 26 – Patterns
Teacher check

Page 27 – Covering Shapes
Teacher check

Page 28 – Lines
V = eleven, approx. 49 cm
S= twelve, approx. 38.5 cm
H = ten, approx. 48 cm

Answers

Page 29 – Classifying Shapes – 1
Teacher check

Page 30 – Classifying Shapes – 2
1. E, F, G
2. A, B, F, G
3. A, C, E, H, I, J
4. L
5. A, B, C, D, K, L
6. A, B, C, D, E, F, G, K, L
7. C, D, E, H, I, J, K, L
8. A, B, F
9. A, B, C, D, F, G, H, I, J
10. H, J

Page 31 – Features of 2-D Shapes
a = square: v – four, e – four
b = rectangle: v – four, e – four
c = triangle: v – three, e – three
d = pentagon: v – five, e – five
e = hexagon: v – six, e – six
Teacher check

Page 32 – Guessing 2-D Shapes
Teacher check

Page 33 – Features of 3-D Shapes
1. A= eight, six,twelve;
   B = six, five, nine;
   C= twelve, eight, eighteen;
   D= six, six, ten
2. Teacher check

Page 34 – Guessing 3-D Shapes
Teacher check

Page 35 – Square Puzzle
Teacher check

Page 36 – Triangle Puzzle
Teacher check

Page 37 – Tangram Pictures – 1
Teacher check

Page 38 – Tangram Pictures – 2
Teacher check

Page 39 – Pentominoes
Teacher check

Page 40 – Looking at Cubes
1. A = eight; B = twenty-seven;
   C = sixty-four
   (all others are the same – faces = six,
   corners = eight, edges = twelve)

Page 41 – Building with Cubes
1. A = five, twelve; B = five, twenty
   C = eight, eight; D = four, eight
2. C
3. D
4. B

Page 42 – Objects and Shapes
1. six squares
2. four triangles and one square
3. six rectangles and two hexagons
4. two triangles and three rectangles

Page 43 – Building Objects
Teacher check

Page 44 – Constructing Objects
1. A = six, nine; B = eight, twelve;
   C = twelve, eighteen; D = five, eight
2. Yes

Page 45 – Tessellations
Teacher check

Page 46 – Symmetry
Teacher check